ISBN 978-1-333-93207-7
PIBN 10745596

This book is a reproduction of an important historical work. Forgotten Books uses state-of-the-art technology to digitally reconstruct the work, preserving the original format whilst repairing imperfections present in the aged copy. In rare cases, an imperfection in the original, such as a blemish or missing page, may be replicated in our edition. We do, however, repair the vast majority of imperfections successfully; any imperfections that remain are intentionally left to preserve the state of such historical works.

1 MONTH OF
FREE
READING

at
www.ForgottenBooks.com

By purchasing this book you are eligible for one month membership to ForgottenBooks.com, giving you unlimited access to our entire collection of over 700,000 titles via our web site and mobile apps.

To claim your free month visit:
www.forgottenbooks.com/free745596

English
Français
Deutsche
Italiano
Español
Português

www.forgottenbooks.com

Mythology Photography **Fiction**
Fishing Christianity **Art** Cooking
Essays Buddhism Freemasonry
Medicine **Biology** Music **Ancient
Egypt** Evolution Carpentry Physics
Dance Geology **Mathematics** Fitness
Shakespeare **Folklore** Yoga Marketing
Confidence Immortality Biographies
Poetry **Psychology** Witchcraft
Electronics Chemistry History **Law**
Accounting **Philosophy** Anthropology
Alchemy Drama Quantum Mechanics
Atheism Sexual Health **Ancient History**
Entrepreneurship Languages Sport
Paleontology Needlework Islam
Metaphysics Investment Archaeology
Parenting Statistics Criminology
Motivational

OXYGEN CONTENT BY THE LEDEBUR METHOD OF ACID BESSEMER STEELS DEOXIDIZED IN VARIOUS WAYS

By J. R. Cain and Earl Pettijohn

CONTENTS

I. INTRODUCTION

The present work is an attempt to throw some light on the relationship, if any, which exists between the method of deoxidation of a steel and the oxygen content as determined by the Ledebur method when carried out as described by Cain and Pettijohn.[1] Incidentally, some other data related to deoxidation problems are given.. The work is incomplete, in the sense that it is not, and does not purport to be, a systematic study of this question, since it represents only (1) modified commercial practice at one plant for the production of acid Bessemer steel, and (2) an experimental deoxidation practice developed for the study of this particular phase of steel manufacture. In view of the great variations in commercial deoxidation practice at different works and with different processes of steel manufacture, this present investigation should be regarded as no more than introductory to the subject.

The data in this paper relate to two sets of steel ingots, the one made during deoxidation tests carried out at the Watertown Arsenal in 1915 and the other made during a cooperative investigation between this Bureau and the Bethlehem Steel Co.'s plant at Sparrows Point, Md., in 1917. The first-named investigation was described by Boylston[2] and the original paper should be consulted. Only so much of the data from Boylston's paper is included here as is considered necessary for correlation with data obtained in the present investigation.

---

[1] Bureau of Standards Technologic Paper No. 118.
[2] Investigation of the Relative Merits of Various Agents for the Deoxidation of Steel. Carnegie Scholarship Memoirs, 7, pp. 102–131, 131–171, 1916.

126112°—19

II. DEOXIDATION TESTS MADE AT WATERTOWN ARSENAL

Boylston's work was carried out in two phases. The first part was a preliminary investigation made by adding to Tropenas converter steel varying proportions of the deoxidizers tested, namely, commercial grades of ferrosilicon, ferromanganese, aluminum, ferrotitanium, and ferrocarbon titanium. The additions were made in a preheated ladle, the metal held for a short time in the ladle after the addition, and then poured into ingot molds which were made of sand. The object of these preliminary experiments was to ascertain the best proportion of deoxidizer under each set of the conditions described in the original paper. The criterion used in judging the effect of the deoxidizer was the relative density of the ingots, which was roughly determined for the whole ingot (these were 20 inches long by 6 inches diameter) by weighing first in air and then in water.

The best proportions of each deoxidizer as fixed by the preliminary experiments were then used in deoxidizing a heat of steel made in a Tropenas converter and having the composition shown in the fourth horizontal column of Table 2.

Table 1 gives the principal details of metallurgical interest concerning this heat, and Table 2 the analysis of the materials used in making it.

TABLE 1.—Details of Heat No. 4458 Made at Watertown Arsenal

	Time of operations		
	Hours	Minutes	Seconds
Metal into converter	2	9	0
	2	12	0
Started to blow	2	12	30
Blow ended (7 seconds overblown)	2	48	40
Converter to 2-ton lad'e	3	0	0

Materials	Pounds
Weight of cupola metal added to metal in converter	425
Weight of ferromanganese added (77 per cent Mn)	45
Weight of ferrosilicon added (56 per cent Si)	7
Weight of converter metal	5092
Weight of steel and additions	5569

TABLE 2.—Chemical Composition of Raw Materials and Steel Used at Watertown Arsenal

Material	C	Mn	Si	S	P
	Per cent	Per cent	Per cent	Per cent	Per cent
Cupola iron	3.40	0.47	1.650	0.047	0.035
Converter metal	.08	Trace	.016		
Converter metal plus additions (calculated)	.387	.68	.218	.054	.040
Same, by analysis	.37	.51	.186	.057	.037
Ferromanganese	6.55	77.00	.908	.024	.185
Ferrosilicon	.16		56.000		

Table 3 gives the compositions of the various deoxidizers used in the investigation and the manner of using these is shown in Table 4.

TABLE 3.—Chemical Composition of Deoxidizers Used at Watertown Arsenal

Deoxidizer	C	Mn	Si	S	P	Al	Ti	Fe
	Per cent	Per cent	Per cent	Per cent	Per cent	Per cent	Per cent	Per cent
Ferromanganese	6.55	77.00	0.908	0.024	0.185			
Ferrosilicon	.16		56.00					
Ferrocarbon titanium	5.92		2.10			None	14.32	
Ferrotitanium C-free			1.30			7.61	24.00	
Aluminum			.87			a 93.08		0.05

a By difference.

TABLE 4.—Time of Pouring Ingots and Details as to Use of Deoxidizers

	Ingot	Ingot	Ingot	Ingot	Ingot
	1, 2, 3	6, 5, 4	7, 8, 9	12, 11, 10	13, 14, 15
	h. m. s.	h. m. s.	h. m. s.	h. m. s.	h. m. s.
Metal into ladle	3 5 50	3 16 55	3 7 30	3 14 48	3 8 55
Deoxidizer added	3 5 56	3 16 56	3 7 30	3 14 48	3 9 00
Finished pouring into ladle	3 6 25	3 17 00	3 7 57	3 15 10	3 9 05
Metal poured into first ingot	3 09 20	3 19 45	3 10 25	3 17 14	3 11 35
Third ingot finished	3 10 56	3 21 20	3 11 34	3 18 31	3 13 00
Deoxidizer held in ladle	0 2 55	0 2 45	0 2 28	0 2 04	0 2 30
Weight of deoxidizer added (pounds)	6.98	6.275	4.16	7.15	2.02
Weight of steel in ladle (pounds)	1054	1015	1067	1000	1395
Per cent of deoxidizing element added	0.095	0.492	0.093	0.40	0.144
Kind of deoxidizer added	Fe-C-Ti	Fe-Mn	a Fe-Ti	Fe-Si	Ae
Per cent of active element in deoxidizer	14.32	79.65	24.0	56.0	98.0
Ounces of deoxidizer (calculated) per ton of steel	212	198	125	229	46
Ounces of deoxidizer element (calculated) per ton of steel	30.4	157.7	30.0	128	45.6

NOTE.—The additions were all made in small preheated ladles each of which had the approximate capacity of the ingot molds used; the steel after addition of the deoxidizer was held in the ladles for the times shown in Table 4, before pouring into sand molds.

a Carbon free.

Table 5 gives the determinations by the present authors of the Ledebur oxygen content of the ingots (two missing) described in Table 4.

Table 6 gives mechanical tests of annealed test pieces taken from these ingots.

TABLE 6.—Tensile Tests of Annealed Forged Steel Specimens (Made at Watertown Arsenal)[a]

Ingot	Deoxidizer	Elastic limit	Tensile strength	Elongation in 2 inches	Contraction of area
		Lbs./in.²	Lbs./in.²	Per cent	Per cent
1F		47 500	82 500	27.5	49.1
2F	Ferrocarbon titanium	52 000	82 000	29.0	51.9
3F		51 000	81 500	28.5	51.9
5F		48 000	83 500	25.5	46.2
6F	Ferromanganese	49 000	79 000	30.5	54.6
7F		46 000	77 000	31.0	57.2
8F	Carbon-free ferrotitanium	48 500	78 000	31.5	57.2
9F		50 500	84 000	26.5	46.2
11F		52 500	82 500	29.0	54.6
12F	Ferrosilicon	53 500	84 500	28.5	51.9
13F		47 000	78 500	29.5	57.2
14F	Aluminum	48 500	78 000	31.0	54.6
15F		44 500	79 500	29.5	51.9

[a] The test bars were prepared from pieces about 2 inches square by 7 inches long cut from the top of each forged ingot (forged to 2 inches round) after a 3-inch top discard. They were annealed by heating at 850° in a Semi-Muffle furnace for one hour and cooled in air; they were then reheated at 850° for one-half hour and cooled in air; finally, they were heated a third time at 650° for one-half hour and cooled in air.

From an inspection of Tables 5 and 6 it is evident that the marked variations in deoxidation treatment of the same steel, as detailed in text and tables, has had but little effect on the oxygen content of the various ingots as determined by the Ledebur method; some difference in mechanical properties is evident, but not so marked as to be considered highly important, except possibly with reference to the determinations of elastic limit.

FIG. 1.—*Ladle test ingots from heat 16600, deoxidized with spiegel.* (*Bottom ingot is blown metal*)

FIG. 2.—*Ladle test ingots from heat 16602, deoxidized with ferrosilicon.* (*Bottom ingot is blown metal*)

FIG. 3.—Ladle *test ingots from heat 16657*, deoxidized *in the ladle with aluminum.*
(*Bottom ingot is blown metal*)

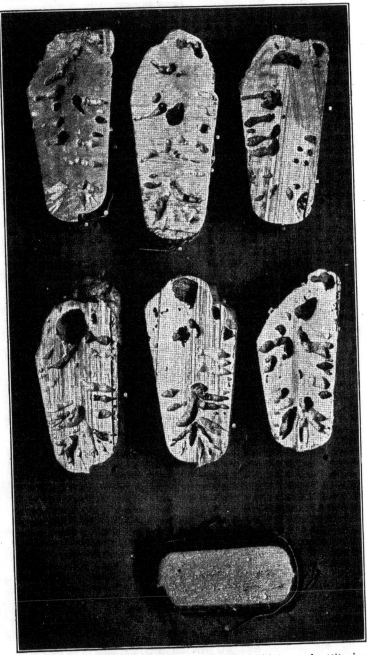

FIG. 4.—Ladle *test ingots from heat 16661, deoxidized with ferrocarbon titanium.*
(Bottom ingot is blown metal)

III. DEOXIDATION TESTS MADE AT THE BETHLEHEM STEEL CO. PLANT AT SPARROWS POINT, MD.

The tests at Sparrows Point were made at the Bessemer (acid) plant by certain variations in the regular commercial practice used here for making Bessemer steel. This practice consists essentially in charging into the converter molten pig iron taken from a mixer, together with a certain proportion of scrap, blowing the mixture in the converter until carbon, manganese, and silicon are removed practically completely; then deoxidizing, recarburizing, and adjusting residual manganese content by the addition of molten spiegel. The details concerning the various heats used in making these tests, so far as available, are given in Tables 7 and 8.

TABLE 7.—Analytical Details of Experiments Made at Bethlehem Steel Co.

Heat	Mixture blown in converter		Deoxidizer		Blown metal		Ladle test of heat					Ladle tests of individual ingots		
	Si	Mn	Si	Mn	C	Mn	C	Mn	Si	P	S		C	Mn
	P. ct.	P. ct.	P. ct.	P. ct.	P. ct.	P. ct.	P.ct.	P. ct.	P. ct.	P. ct.	P. ct.	P.ct.	P. ct.	P. ct.
16600	0.77	1.44	1.37	14.58	0.056	Nil	0.36	0.75	0.05	0.044	0.024	1	0.356	0.75
												2	.358	.75
												3	.358	.75
												4	.372	.75
												5	.362	.75
												6	.366	.76
16602	1.25	.93	a 1.50 / b 50.00	a 14.64	.038	Nil	.38	.72	.05	.086	.042	1	.372	.72
												2	.374	.73
												3	.382	.72
												4	.380	.71
												5	.392	.70
												6	.376	.73
16657	.43	.62	a 1.39 / c 99	a 15.50	.064	Nil	.39	.99	.09	.060	.046	1	.404	1.01
												2	.398	1.00
												3	.390	.97
												4	.392	.98
												5	.410	.97
												6	.380	.98
16661	1.55	1.68	a 1.39 / d 15.50	a 15.38	.092	.08	.45	.95	.05	.092	.016	1	.432	.80
												2	.434	.81
												3	.468	.85
												4	.448	.82
												5	.466	.74
												6	.444	.76

a Spiegel.   c Aluminum.
b Ferrosilicon.   d Ferrotitanium.

TABLE 8.—Metallurgical Details of Heats Made at Bethlehem Steel Co.

Heat	Weight			Time held after adding deoxidizer	Time of heating ingots before rolling
	Spiegel	Mixer metal	Scrap		
	Pounds	Pounds	Pounds	m.	h. m.
16600..	3170	37 000	3500	0	1 45
16602..	3600	42 000	3500	2	1 55
16657..	3560	37 000	3000	2	1 45
16661...,....	3560	38 000	3500	2	1 50

NOTE.—Size of ingots 21 by 23 by 64 inches; weight of ingots 6800 pounds.

Heat 16600 was made as usual, being deoxidized with spiegel; to each ingot of this heat 5 ounces of stick aluminum was added as the metal flowed into the mold. Heat 16602 was deoxidized with ferrosilicon added in the ladle after the spiegel addition; heat 16657 with aluminum in the ladle after the spiegel; and heat 16661 with ferrocarbon titanium (15.5 per cent Ti) in the ladle after the spiegel. Every other ingot of the last three heats received aluminum treatment in the mold in the manner described for heat 16600.

All the ingots were rolled into rails after 10 per cent top and 3 per cent bottom discard and the Bureau was supplied with the

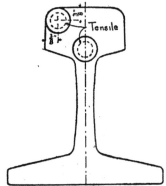

front ends of the first rail from each ingot; one ladle-test ingot per large ingot was also taken at the time when each large ingot was one-half poured; these test ingots were poured from a small spoon which was filled by holding it beneath the stream of metal flowing from the ladle when the stopper was nearly closed. Photographs of the faces of the test ingots on the medial longitudinal plane are shown in Figs.

FIG. 5.—*Diagram showing positions* 1, 2, 3, and 4.
of test bars in rail ends.   For making the oxygen determinations chips were milled under oil (as described in the cited paper by Cain and Pettijohn) from each split-test ingot and from some of the corresponding rail ends. Oxygen was also determined in the blown metal for each heat. For additional information test bars were machined from the rail ends and their mechanical properties determined. This series of experiments was expected to yield information, (1) as to variation

of Ledebur oxygen content corresponding to the different deoxidizers used under comparable conditions;- (2) the effect on Ledebur oxygen content of further aluminum treatment in the molds; (3) the combined effect, if any, of preheating for rolling and of rolling on Ledebur oxygen content; (4) the physical properties corresponding to the different methods of deoxidation, other factors being approximately the same; and (5) the change in these physical properties caused by the aluminum treatment in the molds.

Table 8 gives the Ledebur oxygen determinations on these samples and some nitrogen determinations. Table 9 gives the physical properties of the test bars. Before machining the test bars all the rail ends were put into the same physical condition by heating for three-fourths hour at 800° in a muffle furnace and cooling in the furnace.[3]

---

[3] Some slight change in oxygen content may have been caused by this treatment. However, it would be the same for all the rails and does not affect the comparison of deoxidizers which this study involves.

TABLE 9.—Determination of Nitrogen and Oxygen in Steels Made by the Bethlehem Steel Co.

Heat	Deoxidizer used	Ingot (order in which poured)	Treatment in mold	Oxygen— In ladle-test ingot	Oxygen— In corresponding rail	Nitrogen a
				Per cent	Per cent	Per cent
16600	Spiegel	1	5 ounces Al added	0.021	0.018	
		2	....do	.028	.019	0.0135
		3	....do	.008	.017	.0135
		4	....do	.020		.0128
		5	....do	.006		.0145
		6	....do	.016	.019	.0127
		Blown metal	No Al added	.023		
16602	Ferrosilicon	1	5 ounces Al added	.009		
		2	No Al added	.017	.016	
		3	5 ounces Al added	.011		.0117
		4	No Al added	.004	.012	.0097
		5	5 ounces Al added	.010		
		6	No Al added	.011	.010	
		Blown metal	....do	.024		
16657	Aluminum	1	5 ounces Al added	.017		.0128
		2	No Al added	.015	.018	.0104
		3	5 ounces Al added	.014	.015	
		4	No Al added		.016	
		5	5 ounces Al added	.013	.016	
		6	No Al added	.014	.015	
		Blown metal	....do	.027		
16661	Ferrotitanium	1	5 ounces Al added	.018	.017	
		2	No Al added	.015	.011	
		3	5 ounces Al added	.012	.018	.0151
		4	No Al added	.006	(b)	.0127
		5	5 ounces Al added	.009	.012	
		6	No Al added	.004		
		Blown metal	..... do	.014		

a The nitrogen determinations were made by Louis Jordan, of this Bureau, using Allen's method (Jour. Iron and Steel Inst., 1880, p. 181), with modifications, which will be described in a subsequent paper. This method gives "combined" nitrogen; i. e., that present as nitride. The nature of the particular nitride or nitrides usually present in steel is being investigated at the Bureau.

b Not determined.

IV. GENERAL DISCUSSION

A study of the results given in Tables 5 and 9 discloses no relationship between the Ledebur oxygen content and (1) the various methods of deoxidation used in the ladle; (2) the aluminum treatment in the molds; and (3) the effect of reheating for rolling and of rolling. Nevertheless, the steels in each case had physical characteristics varying somewhat with the deoxidation method, as shown in Tables 6 and 10. These differences are not completely accounted for by other factors, such as heat treatment (which was identical for the steels of each test) or by chemical composition, which was identical for all the ingots made at Watertown, but only approximately so for the steels made at Sparrows Point. Evidently, then, the Ledebur method does not measure anything which indicates efficiency of deoxidation in such steels. Some of the reasons for this are described by the authors in their paper on the Ledebur method (loc. cit.). A probable explanation is that the slags produced during deoxidation (e. g., ferrous silicates, ferrous aluminates, ferrous manganous silicates, ferrous titanates, etc.) and not separated from the steel before solidification are responsible for some of the differences in physical properties shown; such slags, in general, do not yield their oxygen to the Ledebur method, and in this regard the method fails. As shown in the cited paper by the present authors, the Ledebur method can give but little information on the gas content of steel after deoxidation, and in this respect, also, it affords no help in studying the efficiency of deoxidizers, which are really degasifiers.

TABLE 10.—Mechanical Tests of Rail Ends

[The (a) and (b) refer to the position in the rail from which the test bars were taken, as shown in Fig. 5]

Deoxidizer; treatment of ingots	Yield point	Ultimate strength	Elongation in 2 inches	Reduction in area
	Lbs./in.^{2}	Lbs./in.^{2}	Per cent	Per cent
Heat No. 16600 (spiegel):				
5 ounces A1 in mold	(a) 56 500	97 000	24.5	46.5
	(b) 57 200	97 700	....	46.7
Do	(a) 58 000	97 000	26.0	45.0
	(b) 59 800	98 500	23.5	44.2
Do	(a) 56 500	96 500	26.0	47.5
	(b) 58 000	98 750	24.0	42.5
Heat No. 16602 (ferrosilicon):				
5 ounces A1 in mold	(a) 60 000	103 000	23.5	45.0
	(b) 61 500	102 750	25.0	45.5
No A1 in mold	(a) 61 250	100 250	23.0	44.5
	(b) 58 250	100 250	23.5	45.5
5 ounces A1 in mold	(a) 57 500	100 250	25.0	45.5
	(b) 61 200	100 200	26.0	45.3
No A1 in mold	(a) 58 250	98 000	23.0	43.5
	(b) 61 300	100 800	22.5	41.2
Heat No. 16657:				
Aluminum in ladle	(a) 61 500	103 750	24.0	44.5
No A1 in mold	(b) 59 000	102 500	25.5	45.0
5 ounces A1 in mold	(a) 59 250	102 500	23.0	43.0
	(b) 58 500	102 000	23.0	43.5
No A1 in mold	(a) 58 500	106 750	23.5	41.5
	(b) 61 000	106 500	23.0	42.0
5 ounces A1 in mold	(a) 58 250	102 750	24.0	43.5
	(b) 61 800	103 600	23.5	43.2
Heat No. 16661 (ferrotitanium):				
5 ounces A1 in mold	(a) 61 750	113 500	18.0	36.0
	(b) 61 900	113 500	19.5	36.3
No A1 in mold	(a)	108 800	16.5	29.6
	(b) 64 750	122 500	16.5	27.5
5 ounces A1 in mold	(a) 59 500	107 500	12.0	15.5
	(b) 62 250	112 000	18.5	37.0
No A1 in mold	(a) 60 000	105 750	7.5	10.0
	(b) 59 250	108 250	18.5	35.5
Averages:				
16600	57 600	97 800	24.8	45.1
16602	59 900	100 700	23.9	44.5
16657	59 700	103 800	23.7	43.3
16661	61 300	111 500	15.9	28.5

The slight variations in the results for oxygen on the various successive ladle-test ingots of a heat are probably due to segregation; the most unsound test ingots (Figs. 1 and 4) representing heats deoxidized with spiegel and ferrotitanium, respectively, are the worst in this regard; whereas the soundest ones (Figs. 2 and 3), deoxidized, respectively, with ferrosilicon and aluminum, show the least oxygen variation in the successive ladle-test ingots.

The peculiar results obtained by the use of ferrotitanium in the Bethlehem Steel Co. steels are of particular interest. They indicate segregation of some kind, yet the analyses of successive ingots of this heat (see Table 7) for constituents other that oxygen do not show excessive segregation, although they are not nearly so uniform as the ingots of the other heats. In this connection it is interesting that the successive ingots of the heat deoxidized at Watertown with ferrocarbon titanium were quite uniform as regards physical properties, as shown by the data in Boylston's paper. Only 0.003 per cent of residual titanium was found in the ladle-test ingots of heat 16661, so that none of the properties of this steel can be attributed to titanium alloyed with it.

The results communicated by Boylston for gas content (loc. cit.) of the various ingots made during the test at Watertown could not be used by him to throw any light on the efficiency of the various deoxidizers. These results, however, did not represent the total gas content obtained by melting the metal, evacuating and measuring the released gases; instead they were obtained by heating the steel samples to 1000° for 30 hours in vacuno. Just what proportion of the total gas present was obtained in this way is not brought out in the paper, and hence the interpretation of the results is difficult. Work is now in progress at this Bureau on methods for determining gas in steel by melting in vacuno.

The nitrogen determinations shown in the last column of Table 9 are given in this paper for three reasons: First, to supplement the determinations of gases in these steels made by Boylston, which determinations are reported in his paper. The nitrogen shown in Table 9 is undoubtedly to be added to Boylston's results, since the method he used would probably not determine this. Second, to throw some light on the special properties claimed for titanium as a remover of nitrogen in steels. As far as the evidence of this paper goes—and it must be regarded as very incomplete—no more nitrogen is removed by titanium than by the other deoxidizers. Third, to ascertain whether or not aluminum, by the formation of aluminum nitrides, facilitates removal of nitrogen. The results here, also, are negative, but again the same caution should be exercised in interpretation of data.

The authors acknowledge the courtesy of officials of the Watertown Arsenal and of Prof. Boylston for opportunity to cooperate in their investigation; and of the Bethlehem Steel Co., and particularly F. F. Lines, who made many helpful suggestions and facilitated the work in every way possible. They are indebted to the division of engineering and structural materials of the Bureau for the results of Table 10.

V. SUMMARY

1. The Ledebur method for oxygen did not indicate significant differences in oxygen content in steels with nearly identical chemical composition and heat treatment but having different deoxidation treatments.

2. No differences in nitride nitrogen were shown for such steels.

3. The work of Boylston cited in this paper showed no distinctive differences in the gas content of such steels as obtained by heating in vacuuo to 1000° for 30 hours.

4. Judged by the evidence of this paper and that of Boylston, the three chemical methods just named—being those much used heretofore—are inadequate for the study of deoxidation.

5. Mechanical tests show some differences in quality of the steels according to deoxidation treatment, but these differences are not marked and are masked somewhat by other factors.

WASHINGTON, December 23, 1918.

CPSIA information can be obtained
at www.ICGtesting.com
Printed in the USA
LVHW080518271118
598291LV00012BA/1110/P